BACKYARD CHICKENS

Join the Fun of Raising Chickens, Coop Building and Eating Fresh Eggs (Hint: Keep Your Girls Happy!)

© **Copyright 2020 by Rhea Margrave – All rights reserved.**

In no way is it legal to reproduce, duplicate, or transmit any part of this document in either electronic means or in printed format. Recording of this publication is strictly prohibited and any storage of this document is not allowed unless with written permission from the publisher.

The information provided herein is stated to be truthful and consistent, in that any liability, in terms of inattention or otherwise, by any usage or abuse of any policies, processes, or directions contained within is the solitary and utter responsibility of the recipient reader. Under no circumstances will any legal responsibility or blame be held against the author for any reparation, damages, or monetary loss due to the information herein, either directly or indirectly.

The information herein is offered for informational purposes solely, and is universal as so. The presentation of the information is without contract or any type of guarantee assurance.

Medical Disclaimer: The ideas and suggestions contained in this book are not intended as a substitute for consulting with your physician. All matters regarding your health require medical supervision.

Legal Disclaimer: all photos used in this book are either owned by the author, licensed for commercial use to the author, or in the public domain.

By The Same Author

Table of Contents

Introduction..8

1. Why Keep Chickens in Your Backyard?...............10
    Raising Chickens Is Cheaper Than You May Think
    Fresh Eggs. *Every* Day!
    Chickens are Intelligent and Caring
    Hanging Out With Your Chickens Helps
    Reduce Stress
    Free Manure
    Chickens are a Great Pest Deterrent
    Chickens Help You Dispose of Your Leftovers

2. How to Select the Right Chicken Breed...............20
    Show Chickens
    Dual Purpose Chickens
    Egg Producers
    Meat   Chickens
    Space and Size
    Colored Eggs
    Plumage
    Climate

3. Should You Build a Coop, or Buy One?...............34
    Decide How Many Chickens You Want to Keep
    Size *Does* Matter
    DIY vs. Pre-Made Coop

4. What Will We Feed Them?...............................42

How Much Will My Chickens Eat?
Food Management
Growing Ration
How to Feed Your Birds on Range
Set a Good Table for Your Layers
Give Them Fresh and Clean Water, *Always*

5. How to Keep Your Chickens Healthy..................52
Avian Influenza
How to Prevent Infection
Is my Chicken Sick? – Signs and Symptoms of Disease
How to Check for Parasites

6: How to Keep Your Chickens Safe........................62
Your Chicken Coop: McDonald's for Predators
Chicken Snakes
Possums
Owls
Red Tail Hawks
Rats
Weasels
Coyotes
Bobcats
Raccoons
Foxes
Dogs
How to Identify the Predator
How to Keep Your Flock Safe From Predators

7. Incubating and Hatching Chicks........................74
The Nanny Method

Artificial Incubation

8. Challenges: Neighbors and The Laws................82
United States
Australia
Canada
United Kingdom
Rest of Europe

9. A Few Last Tips...88
Giving Your Chickens Greens Year-Round
Put on Aprons to Stop Picking
How to Prevent Water From Freezing in the Bowl
Dealing with Roosters
A Final Thought on Predators

Final Words..94

BONUS: CHICKEN JOKES!....................................96

Did You Like This Book?..104

About The Author...106

Introduction

Thank you for taking the time to buy this book: *'Backyard Chickens for Beginners: Join the Fun of Raising Chickens, Coop Building and Delicious Fresh Eggs (Hint: Keep Your Girls Happy!).'*

This book covers the topic of raising chickens in your backyard, and will teach you the steps you need to take to be successful and keep your girls happy.

At the completion of this book you will have a good understanding of:

- what kind of chickens you want to keep in your backyard
- what kind of coop you will need and how you will need to furnish it
- how to increase your flock
- how to feed them
- how to protect them from predators, and
- things to do in the extreme heat and cold to make the chickens more comfortable and less stressed.

The first benefit you will think of is that you will have a fresh supply of eggs. But chickens are also famous for getting rid of all kinds of bugs and pests. They are very helpful like that. Any weeds you pull out of your garden or around your flowers, they will be happy to eat them for you too. By eating all the extra green garden delights, the yolks of their eggs you will find become more orange by the day.

One thing that few ever think of is sitting in your backyard and hearing the chickens just talk to each other: it is so relaxing after a day at work! They have a soft clucking sound as they go about their business. It is quite soothing to the soul, and if you don't watch out, it can put you right to sleep in that garden swing.

Follow me as you start this book to find out about the magic of having chickens in your backyard. I think you will be pleasantly surprised...

1. Why Keep Chickens in Your Backyard?

KEY TAKEAWAY: *This chapter will give you an idea of the benefits of having chickens in your backyard. It will answer some of the basic questions most people have about chickens in their backyard. It will reveal to you the pros and cons of having chickens in your backyard.*

Backyard chickens are chickens that are kept in an urban setting, usually in one's backyard.

There are many out there that think that if you get chickens that is one more thing to be taking care of at home. But, not so if you manage and set things up the right way.

Raising chickens is very rewarding and a wonderful way to introduce your children to some early farm chores. As a child of about ten, my grandfather purchased twelve chickens for me so that they could lay eggs and I could sell them as a way for me to make money to put back for college. He was teaching me early on how to be frugal.

Chickens make wonderful pets. They are extremely social and very nosey as to what you are doing in your backyard all the time. Chickens have become so popular in Hollywood that a lot of the stars have them for pets in their homes and have designer clothes made for them. With little chicken diapers made to match!

One of the first questions many people ask when it comes to raise chickens is if it is hard and expensive. Worry not: raising a chicken is a breeze. The initial start-up costs some money, but the upkeep afterward is fairly inexpensive.

Let's start with some basics and go from there so you can figure out what will work for you.

***

Raising Chickens Is Cheaper Than You May Think

I feel like having chickens are cheaper than having a dog. And, I have had many dogs, and about every farm animal, you can name growing up on the farm. Their food is cheap, there are all kinds of chicken housing options, ways to water, chicken swings and so forth that can all be made by things you may already have around your house.

I can't think of anything better than going out to the chicken house and gathering eggs, putting them in the basket, coming back in and cooking them for breakfast. The eggs you buy in the store pale in comparison. These come from chickens that have all been housed in a huge long chicken house, with no place to move except their 16 x 16 inch square of mesh and eating, waiting for the time that egg is to pop up and roll down the shaft to be picked up and hauled away to market.

Having chickens in your backyard, and especially if you raise your chicks in the spring, gives you many options. You could either:

- sell the baby chicks
- sell the chickens as they grow
- sell the eggs, or
- process the chicken for the meat to eat yourself

I must confess that last part was a little difficult for me to write as I always wind up making big pets out of the chickens and they wind up living in a chicken nursing home until their time comes to go to that big nest in the sky.

I promise you though, as a child, I helped with many, many chickens as we harvested them for the freezer when they reached the right age.

***

Fresh Eggs. *Every* Day!

A good chicken will produce eggs for you for about two years and sometimes a little longer. You can tell when she starts to slow down. Her eggs will start to get smaller, and we call that her "henopause" years. She will not lay an egg every day, but every now and then she will lay an egg about half the size of what she used to lay as a young hen.

By raising your chickens, you know what is going into them and what is around them. You will know that they are not being subjected to chemicals, pesticides, growth hormones or antibiotics to keep their food fresh longer. You will know the quality of your eggs, a comfortable feeling to have whether you are eating the eggs yourself or are selling them to your neighbors.

These fresh eggs are also better for your health. You will find more than seven times of Vitamin A and Beta Carotene as well as Vitamin E in your eggs over the grocery store eggs.

Your chickens will supply you with 292 mg of Omega 3 while those from the store will give you 0.33 mg. Home grown eggs are lower in saturated fat as well.

***

Chickens are Intelligent and Caring

Chickens are very smart. They also have a great memory and know the difference of over 100, yes, I said 100, human and animal faces. They have feelings too. They dream (don't know how the scientists have proved this, but whatever they say), they mourn for each other, they like to play, and they can feel distress and pain.

They are wonderful mommas. While they are sitting on their eggs waiting for them to hatch, they talk to their babies inside the eggs, and as a worrisome chicken mother will do, the hen will turn her eggs about 50 times a day.

***

Hanging Out With Your Chickens Helps Reduce Stress

There have been studies on the stress reducing effect of kittens and dogs. Just recently, I read about a study showing that cortisol levels dropped appreciably in children who were accompanied by their pet dog when giving a presentation in class.

I am not aware of a similar study on the stress reducing effect of chickens, but let me tell you from personal experience: there is nothing like kicking back at the end of a busy day, with a glass of red Bordeaux wine, and hanging out with the chickens. It is deeply relaxing to tune in to their soft clucking sound as they do their thing.

And they can keep a secret! Seriously, I have shared my deepest feelings and most embarrassing moments with them, and it never came back and bit me in the tail. Although I am pretty sure they gossip about me when I'm not around...

***

Free Manure

A lot of people do not get excited about chicken manure, but our family does. It makes great fertilizer for your flowers and vegetable garden. Put it in the ground below the roots, and you will have some amazing plants that your neighbors will be wondering how you raised such large plants. You can also use your eggshells, but those are also good to give back to your chickens for the calcium which we will talk about later.

The manure great to spread on your lawn, and if you have between five and ten chickens, they should give you about enough fertilizer for your whole yard in a year's time.

***

Chickens are a Great Pest Deterrent

With chickens around, you will not have to buy pesticides if you allow them to free range a little every day. They will eat about any grasshopper, cricket, tick, slug, beetle, tomato worm, flies if they can catch them, and I have even seen them try to peck around on a snake!

They make great gardeners too. You just need to watch out for your ripe tomatoes. They like ripe tomatoes; they like to peck a hole in them, eat a little and go on to another. But they will scratch around and eat the weed seeds that have fallen, helping you with advance weed problems.

***

Chickens Help You Dispose of Your Leftovers

Chickens are great little garbage disposals. They love anything you have left over after a meal. In the winter, when the temperature gets down to freezing, my mother makes sure she cooks them warm food to keep them happy. It seems to work as their production does not slack if she cooks for them. This is a time-honored tradition passed down through the generations to keep your chickens laying well in the winter. They love warm cooked beans or peas.

***

If you must be away for a day or two, I can promise you it is never hard to get a neighbor to watch them, if you promise to let them keep the eggs they gather while you are away.

Don't get me wrong; there will be some initial start-up costs, but all in all, keeping chickens in your backyard is well worth it.

It becomes a great hobby for you and the kids. There are so many kinds of chickens out there to choose from that you just would not believe. And, there are chicken shows, just like dog shows, where you can enter your chickens or your rooster for a prize.

Now, before we move on to chicken coops, let's first take a closer look at the different chicken breeds that you can choose from. Which breed works best for you depends on what you value most. Do you want to keep chickens for show, meat or just their eggs? How you answer that question determines what kind of chickens will enter your coop.

2. How to Select the Right Chicken Breed

KEY TAKEAWAY: *This chapter will go over some of the more common breeds of chickens that you may encounter and help you decide which ones you would like to raise in your backyard. It should leave you curious enough to want to pursue looking up more breeds for yourself as there are so very many to choose from.*

It all depends on what you want to do with your chickens. That sounds odd, right? Do you want them to be:

- Show Chickens
- Dual Purpose Chickens
- Egg Producers, or
- Meat Chickens

Knowing what type of chickens you want to raise is very important. But whichever one you choose, rest assured: all types will bless you with the great but "wonderful" egg.

Other factors that come into play are size, the color of the eggs, plumage and climate.

Let's take a look at each one.

***

Show Chickens

If you want a 'Show Quality' chicken you will want to consider whether you want to 'show' your chicken or sell the eggs for breeding stock. If so, it will be important that your stock starts from a breeder who works hard to maintain the American Poultry Association's "Standard of Perfection." This Standard, which was first published in 1874, classifies and describes things like the standard physical appearance, as well as coloring and temperament for all recognized chicken breeds.

Buying from a hatchery is fine, but keep in mind that their chickens might not hit all the hot buttons for standards when it comes to their exact body type or their plumage pattern. But they are almost always vigorous, healthy, productive and have a very good temper for you that goes so well with a backyard chicken. Your hatchery chicken will also be less of a financial investment.

I have purchased from hatcheries and breeders, and honestly, the hatchery birds came out healthier than the breeder birds in the long run. I found out that the breeder birds were kept day in and day out in a basement. These are NOT the proper conditions for chickens to be living in; they do not receive enough vitamin D3, which is obtained from sunlight, and if not supplemented in the diet they can develop rickets.

***

Dual Purpose Chickens

There are Utility chickens, and if you choose from them, you need to know what you will require from them. Most people who have chickens want them for the eggs, some want meat, and some want eggs and meat both.

Before there was 'factory type chicken farming' there were many who had chickens for eggs and for meat. They wanted to raise chickens that would lay them lots of eggs and chickens that when they got home from church on Sunday, they could go out and chop a chicken's head off and fix it for Sunday dinner.

These were the type of chickens that could roam the yard. They would be a calm type of chicken and just eat what was there for the foraging. If a hen got 'broody' (wanted to lay a clutch of eggs, sit on them and hatch them out and raise them), well, that was fine too. The breeds from that timeframe are still good for today's backyard. So, you might want to try a few of these. They are on the hefty side as chickens. Little plump girls if you will.

These are the main breeds:

- Australorp
- New Hampshire Red
- Rhode Island Red
- Sussex
- Orpington
- Wyandotte
- Plymouth Rock

***

Egg Producers

If your goal is having the girls around to deliver you eggs and you don't need them to be plump for the eating; then you want a streamlined model. They are fine for your backyard, but a little more nervous. Because they are thinner, they can fly better, so you either will need to fix their fence with a covering over the top or be willing to clip only one wing every 4 or 5 months. It is so easy to do.

Clipping the wings stops their ability to balance themselves for the flight so they can never get lift off. Just use a pair of cheap, sharp scissors, the kind you can get at the dollar store. Go out at night while they are on the roost, take your flashlight, take one at a time off the roost, but REMEMBER to clip only one wing. Place that chicken back on the roost and pick up another one and perform the same process all over again until all chickens have been clipped.

Like I said, it is extremely easy; they don't get upset because they are used to seeing you, and it is much cheaper than putting a net or special wiring over the top of the chicken pen.

This kind of chicken will go 'broody' on you easier, which can be a great thing if you want to increase your flock. If you don't want to increase your flock, just reach under her and keep taking the eggs. She will fuzz up and make a funny noise and may even peck at your hand to try and ward you off trying to keep you from taking her eggs. But other than a little-pinched skin, that is about all you will suffer. There won't be any eggs under her to hatch. That would be your birth control if you did not want to increase your flock for that year.

If you should want to increase your flock, bear in mind that if a chicken hatches out say 10-12 babies, it will be about six months from the time they are born before they will start laying eggs. In total, expect to get about two years of her laying eggs for you.

Here is a list of some good egg producers:

- Black Star
- Red Star
- Golden Comment
- Dominique
- Fayoumi
- Campine
- Hamburg
- Leghorn

But there are many others too, you might want to research and decide for yourself.

***

Meat Chickens

If you want chickens just for meat production, then you are going to want one that has large breasts and large thighs and will not waste a lot of their 'calories' laying eggs.

Some of these get to weigh up to seven pounds. These are the best breeds for if you want to keep meat chickens:

- Jersey Giant
- Brahma
- Cornish
- Red Ranger
- Freedom Ranger
- Cornish-Rock Cross

***

Space and Size

It is important to think about your space you have decided on when you pick out your birds. You certainly do not want any overcrowding as this will lead quickly to disease and discord among the fine ladies.

If size and space is your priority, you might want to think about the Bantam Chickens. They have pros and cons to the breed. A plus is that they are very small, about ¼ the size of a regular chicken and weigh anywhere from 1 pound to a 1.5 pounds each. This means they don't take up much space, and in the same coop they will have more roaming space available to them than other breeds. The downside is that they don't lay as many eggs as your typical chicken.

The Bantams have been bred to be more of a show chicken than for laying and certainly not for meat. Bantam eggs are about ½ the size of a regular egg. Regardless, there are a few good reasons that you might want a Banty:

- They don't have to have as much space, they are good for small lots, good if you must keep them locked up all the time.
- They have a calm temperament and are great for children to play with.
- If you want to show your bird, *they* are your show bird.
- They are very broody. If you have a clutch of eggs from a neighbor of some other breed of chicken, put them in a nest and *voilà,* you got a little mother right away.

So here is a list of Bantams in case you want to think about them:

- Booted Bantam
- Bearded D'Anver
- Bearded D'uccle
- Dutch
- Nankin
- Serama

***

Colored Eggs

If you have all kinds of space and can go crazy with all kinds of chickens, you might want to think about chickens that lay different colored eggs. Most of what you will see in the stores is going to be the common white or brown, or different shades of brown.

The coloring of the shell does not affect the taste of the egg. However, I have a lot of the older generation tell me that the brown eggs taste richer. But between you and me: If they were blindfolded I am not sure if they would know the difference…

But in case this is something you particularly care about, here is a list of some breeds that lay colored eggs:

- **Olive** – Olive Eggers
- **Blue** – Easter Eggers, Cream Legbar, Ameraucanas, Araucanas
- **Cream to a light brownish pink** – Old English Game Bantams, Cochins, Dorking, Salmon Faverolles.

***

Plumage

If you are into plumage, you might want to check that out online as well. I think you will be quite surprised at what you find. If your birds are going to be free-range some of the time, bear in mind these few issues about your birds being camouflaged from predators:

- Unless it is snowing all the time, a white bird will stand out easy for any predator.
- Brown or red tones will be better than white or yellow tones.
- If their feathers have a pattern to them like a lacing, speckles, stripes or any other disrupted pattern, this will be harder for a predator to detect them.

- Getting chickens that mimic what surrounds them with their coloring will be the best way to avoid the eyes of predators.

Don't get me wrong though; white chickens clearly have their place. If you are raising that bird for meat, when you pluck a white bird, there will be no dark spots left behind.

There are other plumage options to think about that go beyond pattern and color:

- Some of the breeds have feathers down their legs. It is really pretty, until it gets dirty. Then it is just a nasty mess. It can help them stay warmer in a cold climate to some degree, but if they ever get leg mites, it is a mess getting rid of those. If you don't like shaving your own legs, then you sure won't have any fun treating their legs for mites.
- Silkies and Polish have beautiful plumage, but it can get in the way of their vision. It causes them to not see hawks or owls coming at them. You can keep their head plumage clipped, but it is another job as it grows fast.
- Silkies have more like a fur on them. They are gorgeous. But it looks like fur because the feathers are made differently, and they do not insulate the bird that well. Silkies are wonderful with children, and they want to be

held, they love dress-up, you can put a chicken leash on them, they will follow you around, you can put bows in their hair, and they come in more colors than I could ever list. It is my opinion and mine alone that they are a chicken that needs to live in a warm climate year-round or needs to be a house bird. They are awesome.

***

Climate

Now we need to talk about what kind of climate you live in; and what type of chicken is suitable for your climate. You don't want the girls passing out on you. Let's face it, if they are not heat tolerant, they will go down on their laying, as they are stressed. You must remember, you must keep those girls happy to keep them productive. They don't ask much and give so much in return.

If you need a chicken that needs to hold up to the cold, you are going to need a fat girl. One with a round body, short, thick combs, and wattles that are small. They will be the ones best for cold weather. If their comb is a thin comb, it will frostbite and turn black easily. That must be painful for the chicken.

Your best chickens for cold climates are:

- Wyandotte
- Brahma
- Buckeye
- Chantecler

Now, if you want chickens that can take the heat, we are going to want just the opposite. The girls with the slim bodies, the thin single blade combs.

Here are some good choices for hot weather:

- Naked Neck
- Catalana
- Black Faced White Spanish
- Lakenvelder
- Campines
- Andalusians
- Leghorns

***

There you go: with these tips you should be able to pick the right type of breed for you.

Once you know how many chickens you want, and what size, you will need to give them a place to stay. They can't be free roaming 24/7, with all those predators out there. Remember: make your chickens happy, and you are going to be happy too!

3. Should You Build a Coop, or Buy One?

***KEY TAKEAWAY**: This chapter should give you enough information to help you decide whether to build your own chicken coop or to buy one already made. You will learn what to base your coop size on, as well as what things to consider when deciding on the building itself.*

Building a coop or buying one all depends on if you are a do-it-yourselfer, or are on the other side of the spectrum: someone who just hates messing with putting things together.

I can tell you, if it were left up to me, it would take me months to get a coop built. And I had a shop class in high school! I would be so anxious to get my chickens and so frustrated that I wasn't building fast enough that I would become frazzled beyond belief.

But there are so many out there who could have one built in a day, and I so admire them for that. Deciding on whether to buy or build; there are some things you need to think about before you make up your mind.

***

Decide How Many Chickens You Want to Keep

Think about how many total chickens you want to keep and what size. These are very important questions as it bears on how big your coop will need to be.

If you live in town, you will need to check with your town ordinance as to how many chickens you can have in your backyard. Some Homeowners Associations will not allow chickens in your backyard at all. Some towns will require that the chicken coop is 100 yards away from anything belonging to a business or a home dwelling. Some towns will limit the number of your chickens to three, some to twelve and some have no limit but will not allow a rooster. Just make sure you know about all of this before you start.

***

Size *Does* Matter

After you know how many chickens you want, you need to allow at the very minimum 3-5 square feet of actual floor space for each hen. This will all depend on the size of your full-grown hens.

You will need to think about how much roost space to put up. Allow 8 inches for each hen to be able to roost on at night.

You will need nesting boxes, and the rule of thumb for that will be one nest for about every four hens. You will want your roosting boxes to be about a 16 inch square. And even when you have that many, sometimes they will all still lay all their eggs in the same nest some days. It is a chicken thing, what can I say.

***

DIY vs. Pre-Made Coop

You can buy ready-built chicken coops, but they may not hold as many chickens as you want to work with. Most of them are hard to clean well. You must think of that, because they will need to be cleaned. Where their roosts are located, there will be a lot of poop build up and that will have to be cleaned out on a regular basis to keep the level of ammonia down so that it will not cause your chickens respiratory problems.

One other thing you can do to make your girls extremely happy is to make sure they have a place to 'dust' themselves. That is their way of bathing. You can place a lot of ashes in an area of their chicken yard or some very fine sand or very sandy soil, and they will stand in line waiting for their turn to 'dust' themselves. This also helps them stay free of mites.

You can buy a premade chicken house/coop for anywhere from $200 up to $3,000. Some of the pre-made coops come as they are and you will not be able to customize them in any way. Some will come unassembled, and you will need to do the construction once you get them, so you will need tools.

Honestly, you can probably build a 'large' chicken house for the same money that you would pay for a 'small' or 'tiny' chicken house that is already put together. If you don't have any power tools, you will have to buy them or rent them, so that will be an extra cost.

If you can follow instructions well; you have it made. There are tons of chicken coop plans on the internet for you to choose from; ranging from simple to complex.

Whatever you decide, make sure that the chicken house will be easy to clean and wash out and that it will be easy to replace planks of wood if necessary.

You will need to make sure there is adequate ventilation and that the vents you have are built up high in the coop, so you have a good draft for air flow.

One thing that is of the utmost importance is security for your chickens. Yep, even chickens are not safe. Some predator is always looking for a quick snack during the dark of night. All vents and windows and doors should be covered well with ½ inch welded wire. All latches should be of the kind that are fitted with: guess what?

Predator-proof carabiners, but not eye hooks. Raccoons can lift an eye hook quicker than you can sneeze.

A good place to look online for an entire grouping of chicken coops is 'Urban Coop Company' (**urbancoopcompany.com**). It can give you some ideas, it has plans for if you want to build your own coop, and you can also buy a prefab coop there.

Another place to look at some totally awesome chicken coops made of heavy-duty plastic that looks like it will last for 100 years is Omlet (**omlet.us**). Check out their Eglu coop, for instance. It has the fence and anchors with it and is protected over head. They come in all sizes, starting $675 for the smallest one. It can hold up to ten chickens and has the nest and roosting area built in. You can pick it up, and it has wheels on it to roll it anywhere in your yard that you want. It comes with feeders and waters and all those extras you need. I do not know who came up with this concept, but my hat off to them.

***

My parents bought one of those little wooden storage sheds that looks like a barn. They went into it and added windows for circulation, and then placed a window on its side down by the floor so it was like a sliding patio door with a ramp that led to the outside.

They placed a roost on one side, and their nesting boxes on the other wall. The feeder, they hung from the ceiling. Now, are you ready for this? She uses box fans in the summer and a heat lamp in the winter. Outside, in the summer, she uses sprinklers to keep them cool. She has a very nice chicken run for them with chain link fence panels that are rather tall, and they have worked out great for her. She has some very happy chickens!

No, my mother is perfectly sane. She just loves her chickens. And my father does whatever makes my mother happy. He also happens to be an excellent carpenter. Win! The wooden storage shed made a perfect house for her chickens, and it houses about 24-40 hens very nicely.

Giving your chickens a roof over their head is one way of making them happy. However, there is more to that equation. Can you guess what it is?

4. What Will We Feed Them?

KEY TAKEAWAY: *Feeding is a very important factor in the health of your chickens. This chapter will guide you as to the proper feeding at the different stages of your chickens' lives. It will also inform you on what other things besides chicken food you can and cannot feed your chickens for their optimum health.*

How Much Will My Chickens Eat?

Besides a place to rest their beak at night, chickens also need to eat and drink to stay happy and lay those yummy eggs.

A chick that is normally maturing, like the one you are raising to lay eggs, will probably eat about two pounds of food (starter) in her first six weeks of life. If you have a breed that you are raising; a breed that grows fast, like one that you will use for meat, they will eat almost eight pounds of food (starter) in their first six weeks after hatching. But remember, they will grow extremely fast and only be two months old when harvested.

Your laying hens will consume different amounts. Several factors will affect how much they eat. For instance:

- what type of chicken breed you keep
- how much they forage and find food on their own
- how nutritional the bought food is they eat
- humidity
- if it rains or is windy
- if it is really hot or cold, and
- if they exercise much

Don't forget that you will have all kinds of critters steal food in the dark of night, unless you take it in every night. Those pesky mice and rats love chicken food. If you have a hole big enough, a baby opossum will sneak in and eat it too. And birds will eat you blind! They like to come in just as the sun is about to rise. They will be there in flocks!
You might want to keep records of just how much feed you are using for your girls.

***

Food Management

If you want to make sure your birds are healthy, make sure that the food you have out for them is always fresh. If you have a hanging feeder – my personal favorite as it saves on waste –, fill it only three-fourths full. If you have a trough feeder only two-thirds full and they will try their best to scratch it out. If you do decide on a hanging feeder, keep the lip of it at the level of the girls back.

If you have non-automatic trough feeders, you might need to fill them twice a day. If the food left in it is not clean, you will need to dump out what is in there and refill it. Keep your feeders clean.

If you notice one of your girls is losing weight or does not touch her food, you can be sure that something is not right. It could be because she is molting (which is normal when they lose a large number of feathers at one time), but I have found that if you give them extra greens from the garden or extra cooked food, that they do not lay off their laying that much. Sometimes, if the molting coordinates with extreme temperatures, that will hamper laying more.

It is wise to store your food in a large galvanized can with a good tight lid.

***

Growing Ration

Your local feed store can help you with so much good advice. That is because they are used to new folks coming in and starting to raise their own chickens. They are ready for your questions, and most are very patient when helping you out with what you need. All you need to tell them is how old your chickens are and if they are layers or fryers.
As your chickens age, they will need something called growing ration. It is so much easier to buy your food for your chickens than you buy the separate parts and mixing it yourself.

If your girls are:

- **Six to fourteen weeks old**: you are going to need about seventeen percent of protein in their food mix.
- **Fifteen to twenty weeks old**: now you will need to make sure they have fourteen percent protein in that food mix.
- **Eighteen to twenty weeks old**: at this stage, start pulling away from the growing mash and introduce the laying mash. You are getting to the stage that you are only a month away from them giving you eggs.

If you want, you can give them grain along with their mash. It will help with the overall cost. You can feed grain to your girls as soon as they begin eating growing mash.

***

How to Feed Your Birds on Range

When they forage for their food, they will not be able to get everything they need for their diet to stay healthy and produce an abundance of eggs. What is most often used is either pellets or mash in one feeder and some grain in another.

Some chicken farmers had rather use pellets because they are not as bad to blow out of the feeders.

They recommend about four inches of feeding space per bird. This is so they can all eat at the same time. Other than when they are new chicks (and they all do eat at the same time, and it seems they eat all the time, crawling over each other and on top of each other), I do not agree with this as when they get older, the girls do not all eat at the same time, I can assure you of that.

There is one older black hen in my mother's flock who meets her every day at the entrance gate to the chicken yard. She follows her every day to where the cracked corn is stored (like she is supervising) and then follows my mom out into the chicken yard while Mom sprinkles it out and talks to the girls. Now, I can assure you, this black hen is no longer productive, but she is like the matriarch of the group and I guess she feels she must preside over essential functions such as this. She has been doing this for quite a while now has marbles, and when she is no longer with us, I am sure she will be missed.

***

Set a Good Table for Your Layers

It is so important that your girls have a 'balanced' ration to keep them in tip top shape to be high producers for you. Your best choice is to buy it already mixed at a feed store. Your laying girls need a laying mix with at least fifteen percent protein. You will find that there have already been minerals and vitamins added to this mix.

You are going to find, after your start-up cost, that food is going to be your biggest expense. If you buy your chicken food from a feed store, you should not have to give them any calcium supplements or oyster shells. And, if your hens are not allowed to be on the ground, the food these days even has grit in it.

If you are going to give your girls grain on top of their feed, you will need to give them no more than a pound and a fourth for every ten hens each day.

As I referred to before, table scraps, extra milk, and green stuff from your garden can be used to feed your hens, and that will reduce the food costs. The amount of food you feed should be limited to what your girls can eat in under twenty minutes. They love cabbage, turnips, leafy vegetables, peelings, cauliflower, stale bread, turnips, squash, cucumbers, kale, spinach, lettuce, okra, corn, black eyed peas, and I could go on forever.

You want to avoid some things:

1. **Plants Belonging to the Nightshade Family** – Examples are tomatoes, potatoes, and eggplants if they are in their un-ripened state.
2. **Foods High in Salt** – These can cause salt poisoning.
3. **Citrus** – If they ingest a lot of it, it will cause massive feather plucking.

4. **Onions** – If they eat massive amounts it can cause anemia, jaundice and even death.
5. **Undercooked or Dried Beans** – They have poison in them that is toxic to birds.
6. **Dried Eice** – If they eat it dried, the rice can blow up and cause gut issues.
7. **Skin or Pit of Avocado** – Chickens do not care for them anyway, and it is probably because they can smell the toxins.
8. **Raw Eggs** – This can cause chickens to turn to cannibals and start eating their own eggs.
9. **Chocolate, Sugar, Candy** – Chickens only have like 25-30 taste buds. Sweet, processed foods are bad on their stomach and chocolate is poisonous to chickens.
10. **Apple Seeds** – They contains tiny amounts of cyanide.

It doesn't take much to kill a chicken. For example, if you use laying feed for your growing chicks, it can cause kidney damage. So be careful.

As a rule of thumb, a 25-pound bag of food should last 10 of your hens about ten days, if the food waste is controlled and you are feeding a decent food.

***

Give Them Fresh and Clean Water, *Always*

Make sure that they always have a lot of fresh, clean water to drink. This is very important in keeping a chicken healthy. During the summer, you will need to give them fresh water two or three times a day to keep their water cool for them.

***

If you just give them the basic care of food and shelter, your chickens will pay you back tenfold in eggs, love and cuteness.

Unfortunately, to keep your girls healthy long-term, sleeping and eating is not enough. Like any living being, they can get sick. In the next chapter we will discuss the best ways to prevent your chickens from getting sick, and what to do once they do get sick.

5. How to Keep Your Chickens Healthy

KEY TAKEAWAY: *Within this chapter, you will find a few things to watch for in your chickens to make sure you are ahead of the game if they show signs of getting sick. It will also give you some things to be able to take care of some of the issues encountered if your chickens become infected, as well as a few prophylactic measures in keeping them healthy.*

When it comes to keeping your girls healthy, it all boils down to keeping them fed right and keeping them happy. The rest will just fall into place.

If your girls are not getting the foods they need or they feel stressed for some reason, or they are not staying warm or cool enough, you open the door for disease to take over. You need to keep a watchful eye on all of them and if only one shows up with a sign or symptom, treat it immediately before it can spread to any of the others.

For some, you may think this is a little overkill, but you wouldn't if you walked out some morning and over half of your chickens were dead or dying and the others were showing symptoms.

The Avian Influenza (AI) strain is one of the biggest dangers threatening your chickens. It spreads so easily that I would hate for it to get into any of my birds. They are like family to me.

So, after I have explained what it is, I will share my tactics with you on infection control prevention to keep your girls healthy. These methods also help prevent other diseases.

***

Avian Influenza

AI can be spread easily. Just being in close contact can cause an outbreak. Because of AI, there have been many poultry shows canceled over the past few years due to the fear of this deadly disease.

It has become a risk just going to the feed store. If somebody has a sick bird and doesn't realize yet what is going on and has the AI on their hands. They climb in their car and you know it's on their shoes and probably on the tires of their truck and on their hands.

They get to the feed store, touch the handle outside to open the door of the feed store (BAM, contaminated) ... then they grab a cart (BAM, contaminated) ...

Now here you come to the feed store, you drive onto the lot, walk up the same steps, and open the front door (BAM, contaminated from where his hands had touched it) ... then you go in and get the cart that he just put up and (BAM, contaminated)... Now, you can take it home to your girls.

I know that this sounds unrealistic, but I promise you: it is not. This is called the "chain of infection" whether it be in the human world or the animal world.

***

How to Prevent Infection

Here are some helpful tips to keeping your girls as healthy as possible:

1. **Keep Hand Sanitizer Around**. The pump kinds. Use it frequently. If you shake someone's hand, use it. If you pick

up a bird, use it. You can get the little bottles to carry in your pocket too.

2. **Tires and Shoes**. You need to wash and disinfect before you go back to your home. I am not kidding. If you want, take a pair of boots or other shoes you can bleach when you get home, even better. If you can spray your tires down with bleach before you run through a car wash. Even better.

3. **Some Kind of Tarp**. If you happen to go to a chicken swap, put down a new clean tarp to set your cage on. After the show, roll the tarp up, keeping all the poo rolled up in it. Burn the thing as soon as you can. People have been walking on it, and they have poo on their feet. They could have AI on their feet.

4. **Plastic Tubs**. If you take baby chicks or ducks to a swap, it is usually in a tub. These can also be washed out with bleach on the way home at the car wash.

5. **The Truck**. Always try to take a pickup to a bird show, you can put down the tailgate and set your birds across the back of it with the tailgate down. Use bleach and go by the carwash on the way home.

6. **Full Boxes, Signs, Empty Cages**. When you get to the swap, your cages are empty. The birds have been kept in boxes in the truck in the back seat. Just take out a few of each breed to put in cages for the show. Use small dry

erase board, write what you have, and prices. Birds in the back seat are never exposed to anything. The ones in cages will be sold, or if they must go home, they will get quarantined for two weeks.
7. **Car Wash**. Go through the car wash, bleach tires on the car and bleach tubs before going through the car wash.
8. **Arrive at Home**. Take your boots off and place them in the driveway. Go on in the house, change your clothes, put your dirty clothes in washing machine carefully before touching anything.
9. **Quarantine**. Now, put away all clean birds in the truck. Set upa quarantine area. Put on rubber gloves. Take the birds that need to be placed in quarantine and set them up. Burn the rubber gloves and boxes the quarantined birds were in. Spray bleach on the tires of the truck, on your rubber boots, in the bed of the truck and then rinse off with the hose.

That was a mess, but at least it should not have carried any diseases home to the other birds.

Perdue Farms uses oregano in their chickens' water as it acts as a natural antioxidant for the chickens to keep them healthy. They use thyme in their chickens' veggie feed as it adds more flavor for their chickens and it is strong support for their immune systems. They never use antibiotics for any of their chickens.

Keep your chicken house clean. Do not let their roosting area pile up with manure. It will cause such a terrible ammonia smell that it will cause them respiratory problems.

When you clean your coop, wear a face mask that can be bought at your local Walmart, so you will not be breathing all that dust into your lungs. It is common for avian species of all kinds to carry some diseases in their feces that – once the feces is dry – can become airborne, and you can breathe it into your lungs. So, just in case, take the precaution.

The manure and straw/shavings or whatever you use will make a superior product for your compost pile or for your garden or yard, whichever one you want to use it on. Your chicken poop, your choice.

It is a clever idea after you get it cleaned out to come in and spray the area down with a 1:10 mix of bleach. It takes very little for a kill. Let it set there for about 10 minutes and then when it is dry you are ready to put more straw or shavings back into your girls' house.

If you purchase any new birds to add to your flock, make sure that you keep them separate for 30 days. This is to make sure that they have not been exposed to an illness that they are not showing signs of yet.

This is very important, and I cannot stress this enough. Whatever you do, do not ever share birds with someone, nor share garden or lawn equipment, poultry supplies, and tools with any other bird owner or even your neighbors.

**

Is my Chicken Sick? – Signs and Symptoms of Disease

Here are some signs and symptoms of disease you need to watch for in your chickens:

1. A chicken that was just fine dies suddenly
2. Diarrhea
3. They have quit laying eggs, or their eggs have all gotten soft shelled, or the eggs are misshaped
4. They have started coughing, have a nasal discharge, sneezing and gasping for air
5. They have no appetite and lack of energy
6. Around their eyes and neck, there is swelling of tissues

7. There is the purple coloration of their legs, combs, and their wattles
8. They have drooped wings, muscle tremors and seem depressed
9. They are completely paralyzed, uncoordinated, and have twisting of head and neck

They might show signs of worms, ticks, lice, fleas, and mites.

Earlier in this book, I explained about having a place to dust as it would help in controlling the mites and the lice.

If you have a very large group of chickens, you may even want to consider vaccinating for fowl pox. It is not hard to do. You have the 5 or 6-prong needle device, dip it in the vaccine, pull the chicken's wing out straight, and punch the needles into the thin part of the chicken's wing, close to their body. You sure do not want to get fowlpox in your chickens. It is carried by mosquitoes.

***

How to Check for Parasites

The best way is to get a stool sample and take it to the vet so they can put it under the scope and look for the parasite or the eggs. Usually during warm weather is when this is a problem, knowing when it is worse, you will want to worm after the season has passed.

Sometimes chickens are bad at picking at one another, just an ugly habit, sometimes of just boredom or overcrowding and at times, just because. When chickens start their picking, they usually find one girl in the bunch and pick her till she has almost no feathers left. It can start over the most minute things. Betsy Sue Hen thinks she sees a drop of blood on Curly Kate Hen's head and she pecks at it. When she pecks at it, she draws blood for sure. Then Martha Nosey Hen comes running over to see what Betsy Sue and Curly Kate are doing and sees the blood, then Martha Nosey starts picking, and before you know it, all the girls are there, and they are all picking at poor Curly Kate and almost all her feathers are gone!

The best remedy for this is to cover Curly Kate's bald area with either pine tar, which sticks to that spot so good and the others will not bother her any more, or Vick's vapor rub or its generic version.

I am a believer in the pine tar. I have used it for dipping a feather in it and putting it down the back of chickens' throats when they get a cough. It works great for that too. It works immediately. Sometimes in the summer, after you have cleaned the chicken house and there is dust, the chickens will get a little hacky cough, I have performed this procedure with excellent results.

***

If you apply these tips and tricks, you can be sure that your chickens will live happily ever after.

Or will they?

There is one more major threat that we really need to talk about.

6: How to Keep Your Chickens Safe

KEY TAKEAWAY*: In this chapter, you will be forewarned of all the predators that are lurking in the shadows, ready to pounce and have the delicious chickens you have raised and taken tender loving care of; for their lunch. It will also tell you what you can do to try and prevent this from happening to your flock.*

Your Chicken Coop: McDonald's for Predators

Your chicken pen and house looks like 'McDonald's' to hungry predators. And they prefer a McChicken over anything else on the menu...

No matter how prepared you think you are in taking care of your girls, believe me, some predator you do not want to come in will be able to find its way into the chicken house. Anything and everything from miles around seems to find their way to the chicken house.

Let's take a look at some of the most common predators, and what you can do to keep your chickens safe.

***

Chicken Snakes

One of the most aggravating problems we have had has been very large chicken snakes. These snakes favor eating eggs and small birds over anything else. Imagine finding them in the nest when you go to gather eggs, and all the eggs are gone.

Now that makes for one mad chicken owner, let me tell you.

She can pick that snake up and slam him to the ground, breaking all the eggs in him in nothing flat. He is sorry right then and there that he has eaten all those eggs. There he lays in the nest with ten lumps spaced out the length of his body. The next thing he experiences is a pistol going off right behind his head and a very old yellow lab shaking him until the broken eggs come stringing out of his body. What a way to start the day. But, you just don't mess with someone's chickens. It's like a cardinal rule or something. I honestly do not know of anything to do for snake deterrent. Usually, egg and baby chick eating is what you have to worry about with them.

***

Possums

There are at least five snakes a year that we must go through the drama with. There are also about three possums sited in a year.

I know they are supposed to be predators of a sort. But being a possum lover, I just can't see them bothering a chicken. Now, I know they like to eat chicken food and maybe they have eaten an egg or two, but I don't think they have ever killed a chicken, even though there have been three that suffered the same fate as the snake for being found sleeping in the girls' nest early in the morning. Gee, you wish they had known better.

***

Owls

And the owls, they are sneaky. They know just when to fly in at the right time of night and grab a chicken and have their way with her for meal time. The owl never carries her all the way off, just has part of her and leaves the rest. This puts fear in the rest of the flock for sure. And for a day or two, they may not lay as good.

***

Red Tail Hawks

Then there is the red tail hawk, that evil little devil. Still not sure why the good Lord made that one. He is a sneaky one.

There are so many of them around this year, they have stooped so low as to try and pick off the birds from the bird feeders. They manage to fly in over the top of the chicken house door and it is a narrow space, but they are like a stealth fighter plane and then do their evil deed, and off they go.

***

Rats

All along, I never believed the judge, jury, or executioner about the possum. That the said possums were coming in and killing the chickens for sport and just eating their heads off. It was just too out of character when all the free food was there so handy.

My point was finally proven the morning that the biggest rat in the entire world was found in the chicken house. I am telling you it was as big as a possum; a full-grown possum!!! I am sure it had been the culprit all along. Of course, the judge, jury, and executioner went quickly into action, and he lives no more, and the madness of the chickens without heads has ceased.

Now, the judge and jury, not the executioner, feel certain that on two hens, it was a possum that was the silent killer. And the thought behind this was that whatever it was ate the butt end out of the chickens on two separate nights. Possums have been known to disembowel their hen victims. So maybe, but it is just hard for me to believe as possums are just lazy. Smart, but still lazy.

The executioner has gone so far as to set up two or three trail cameras in the chicken house to take pictures in case something comes in. Leaving them there for two weeks at a time. Nothing happens. Take them down and the next night, evil strikes. But, I think evil in the form of a rat, is gone now.

If you have small chicks you are raising in your hen house, make sure that there are no cracks in the floor. It is a favorite pastime for the rats to reach up through the cracks, pull the chicks legs down through the cracks, and eat their legs off.

***

Weasels

If you have never seen the damage a weasel can do to a hen house, then you just cannot imagine. A weasel can come in and kill every chicken you have in one night. They may not eat one chicken. They just do it for fun. They like to travel in twos or threesomes. I have known of them killing 500 hens in one night.

***

Coyotes

There has been little experience at this farm with raccoons, foxes, weasels, or coyotes. There are two golden labs that watch the place, and if a gnat burps, they will be around to find out why he did. Coyotes try to come in close at times. They will run in packs and will take out every chicken they can. But coyotes will stay away if you turn all the outside lights on around the house. The large dogs bark wildly and soon the coyotes decide to annoy something else.

***

Bobcats

Bobcats will sneak in as well, but they will usually kill only one chicken, pull back the feathers and eat some. They may kill another one and take it with them. For later. Depending on how hungry they are. They will not eat the feathers. They can jump high and more than likely will not try to go under a fence.

***

Raccoons

The raccoon, however, is such a sly little guy. If you have one for a pet, he will unlock his pen, run around all night and come back around 9:00 a.m. and put himself back in pen and lock himself back up. Really? Now that is smart!

There is no lock you can put on a chicken house that he can't figure out unless it is a padlock and you hide the key. The raccoon is brilliant and do not think for one minute he can't figure out what you have done with a few boards and some bungee cords. That is just a fun game to him.

***

Foxes

Surely you have heard the expression 'Sly as a fox'. It is so true when it comes to your chickens.

A fox can dig where they need to and they can dig fast. They kill for food, and they are quick. They will kill everything in the henhouse in one night, and what they can't eat, they will get it carried back to their den to have to eat on for the week ahead. If they must make twenty trips that night, they will do it. But they will carry all of them off and leave a bloody mess for you to find the next morning.

***

Dogs

There may also be dogs in town that are roaming the neighborhood and attack your hens. Just remember that in most towns there are leash laws. If this is the case in your town, their owner will be responsible and must compensate you for the loss and the damage they did to your chickens, as well as to the hen house. Just make sure you catch them at it, so you can identify the dog.

***

How to Identify the Predator

I hope I did not scare you by listing all those predators out there. But you need to know what you are up against.

If disaster strikes and something has happened in the hen house, how do you know which predator is responsible?

By taking a look at the clues. Each predator has its own way of attacking prey. If you put the puzzle pieces together, you will be able to identify the predator. Let's go over a few different scenarios:

1. **Adult hens are missing and you can't find any other clues**: your predator could be several different enemies. An owl, a hawk, a bobcat, a fox, a coyote, or a dog. More than likely it is going to be a fox. But think about your area and your specific situation and your hen house.
2. **Baby chicks are missing, but there no other clues**: your predator is probably a housecat, a raccoon, a rat, or a snake.
3. **Hens are dead but not eaten (sometimes the internal organs have been eaten)**: it is probably a mink or a weasel.

4. **A hen is dead, and only their head is missing**: it is more than likely an owl, a hawk, or a raccoon.
5. **Your hen is only wounded and not dead**: it could be a dog. If you find wounds only on the legs or breast of the young birds, a young possum could be your problem. Now if the hen shows that her intestines have been taken out through her anus, it was probably a weasel.
6. **Only eggs are missing**: you must look at several different predators – crows, blue jays, raccoons, possums, rats, snakes, and skunks.

Be sure and look for tracks around the outside of your pen and in your pen if the ground is right. This might tell you very quickly who the predator is.

***

How to Keep Your Flock Safe From Predators

I am sorry if that was a bit graphic. I hope you will never have to wake up, walk outside and find such a scene.

Here are some tips to hopefully help you in keeping your flock safe:

- Make sure your coop is built to be very sturdy and that the locks and latches are something that a raccoon cannot figure out.
- Keep the chickens area free of any leftovers. This food will be a great attractant for possums and other vermin that you do not want.
- Keep a close eye on your coop and watch for any areas that might have a hole or what looks like something might be going under the fence. Weasels can get through a spot only a ½ inch wide. Some folks put a layer of hardware wire under the floor of their coop.
- Watch when you let your girls go free range. Make sure it is not a time when coyotes will be roaming. Like right after dawn and right before dusk.
- If you have a chicken run, the lower three feet need to be enclosed with hardware wire that is ½ inch. This way those pesky possums and raccoons won't be able to get through.
- Livestock dogs for guardians are very popular, and if the dog is trained to protect your hens, you will have no worries whatsoever.
- If hawks or owls are a problem for you and it is too expensive to cover the top in the wire; try crisscrossing the top with string or wire as this will make it hard for the birds to get down into the pen.

***

The sight of a predator is one, if not the, biggest stressor for a chicken. You will score so many brownie points with your flock if you can protect them from their worst enemies and keep them safe.

Now let's move on and talk about how you can increase the size of your flock.

7. Incubating and Hatching Chicks

KEY TAKEAWAY: *In this chapter, you will find how to increase your flock numbers, whether it be naturally with the use of a mother hen or artificially with the use of an incubator. It will guide you through how to feed them as well as what to feed them with during this time.*

There are two ways to increase the size of your flock:

1. **Natural**, or what I like to call: **the Nanny Method**
2. **Artificial Incubation**

If you have chickens long enough, believe me, you will one day, at some time experience both methods of raising your own chicks. But, I for one, will tell you this: I very much prefer the nanny method!

***

The Nanny Method

The nanny method is where you find a hen that is 'broody' and in the mood to 'sit' on a clutch of eggs.

Give her the eggs; by all means, let her have it. Let her do all the nanny work of fussing over those babies and getting them grown. Give the girl some food and water; so it is close to her. If you plan on using any eggs from anywhere else, say another type of chicken, put those under your girl at dusk. If your girl hatches out 80% of her eggs, she has done a wonderful job. It is a big job if you are doing it yourself with an incubator. She will be so good to turn those eggs 50 times a day, fuss over them and care for them.

Nature has made her to be the perfect incubator. When they are born, she will teach them how to eat, where to get water, and then teach them how to forage for food. She will be their protector from all things, even other chickens. She is no longer a chicken; she has become a lion as the eggs have hatched. She will be with them until they have grown enough to take care of themselves. You will still need to have available for them the starter food, and the grow food at the right ages. But momma chicken/lion will do all the rest for you.

Now, if you want to try out artificial incubation instead, I will share this delightful, but work intensive experience with you so you may enjoy this yourself.

***

Artificial Incubation

When you incubate the artificial method, it will take the same number of days as it does as if it were under the hen. That is 21 days, so if you set your eggs on a Monday, you can bet that in 21 days on a Monday is when they will hatch.

Make sure you have undamaged, even-shaped, and clean eggs to incubate. Set your eggs within a week after laying. You want eggs that have not been washed.

Before putting your eggs in the incubator, go ahead and plug it in and see if the temperature will stay steady. You will need a turkey baster to add water every day, sometimes twice a day to maintain the humidity at the right level. The turkey baster is an absolute must and hard to find at other times of the year besides Thanksgiving and Christmas. The water must be placed in a specific area in the incubator for it to be heated and to maintain the humidity at the right level.

If it is a forced air incubator (my preference) and most of them are, that are made today; the temperature should stay around 99-99.5 degrees Fahrenheit. In the incubators today, the hygrometer and the thermometer come made on the incubator. It is important that on days 1-18 you keep the humidity at 45-50%, and for the last three days it needs to be 65%. Make sure that after day 17, you stop the eggs from turning.

Most of the incubators you can purchase now have the egg turning feature in them also. So, you do not have to make sure you are turning your eggs twice a day. This is something that MUST be done.

Get a marker that is non-toxic and mark each of the eggs with an (x) on one end and (o) on the other. This is to help you know they are turning as they should. The incubators of today will turn them for you, but marking them will let you know for sure. The newer incubators are more like momma. Turning is extremely important and if they are not being turned, that can cause the chick to grow abnormally.

The newer incubators will run you about $40 - $50 or less, depending on where you buy them.

When the eggs start to hatch, it is like watching a miracle. At first, you can't believe it is happening. You see a small hole in an egg and a crack in another. You get to looking and then see another crack and some more holes and realize they are hatching. The chick has used its egg tooth to make that hole. After she has done that it will rest for 3-8 hours. During this rest period, he is letting his lungs learn how to breathe. It will take a few more hours before the chick is free of the shell and when he breaks free, he will be exhausted and lay there and rest for a while. This is all okay and the way nature made it to be.

You know you can't touch them, you must let nature take its course, but it is so hard not to get your hands in there and do something. When they finally come out of the egg, you just cannot imagine how that chick got into that little egg. NO way! This was way cool. You have been waiting for 21 days for this, and it happened.

When your chicks hatch, you will be busy. You will need to dry them off, and you will need to fluff them out before you can put them into the brooder. Don't try to feed them right then. They will live off the yolk they absorbed for about three days.

When you do put them in the brooder, you will need to make water available for them and offer them food after a day or two. For a brooder, I use the smallest kiddie swimming pool I can find and put shavings or shredded paper in the bottom. I place my special waterer in there with the marbles, so they will not fall in the water and drown.

Then I hang a heat lamp from above over them to keep them warm. The temperature should be kept at 95-100°F for the first two weeks and then reduced by 5 degrees (by raising the lamp) until they are one month old. When you put out water, make sure that you put rocks or marbles in the bottom of the whatever you will be using for them to drink from to keep the chicks from drowning.

They will need to be fed 'chick starter' and they need to have feed available at all times. They will be thirsty and hungry and will be eating constantly. They will grow so fast you will not believe it. That little fluff of yellow will soon start to shoot out pin feathers. You will know when you can put them out into your chicken house. If you put them in with other chickens you may have to watch them or put them in a crate or smaller pen for a few days till they all get used to each other so that the bigger hens get used to them and do not chase them.

***

Like I said, I have a personal preference for the nanny method. But by all means, pick the one you like best. Or even better: try them out both, so you will have first hand experience with both methods. Then you will know which one is your absolute favorite.

Now, so far we have assumed that your neighbors are okay with you keeping chickens in the backyard, as long as you give them some of those delicious fresh eggs in the morning. However, that may not be enough. Some people, huh? Therefore, before you get your chickens and coop, double-check that you comply with local regulations, if any.

8. Challenges: Neighbors and The Laws

KEY TAKEAWAY*: This chapter gives a little advice on some of the rules and regulations for housing chickens in your backyard. If you can, and if so: how, when, and where.*

In some parts of the world, a crowing rooster is a welcome sound. But, in other parts of the world, that is not always the case. Especially in the more social parts of the cities, where folks tend to like to sleep in late on the weekends and holidays. They are not happy when a rooster crows to announce the morning sunrise.

***

United States

If you live in the United States, it all depends on the 'zoning' regulations, and the by-laws and the laws of your State, community or town.

If you live out in areas of the country used for farmland, there should not be any problems raising chickens or roosters.

In town or the burbs of the towns, well, that is a totally different story. Your rooster will most likely not be welcomed.

Every town is very different, I can assure you that. Whatever the council comes up with basically. Some even tell you how your fence must be made and where it must be located; to keep up with the looks of the neighborhood.

***

Australia

In Australia, all over the entire country you will find that people are pretty laid back about the rules on chickens if they are well taken care of. Each 'territory' in Australia has different laws about backyard chickens and there are times that your local councils might impose some kind of by-law that will override all else. So, be proactive and find out what the rules are in your area by checking with your Town Hall. Ask for the regulations on "Poultry Keeping on a Small

Scale." If they do not have anything regarding this; it means there are no rules. So, you can do whatever.

Like most other places, if you have a crowing rooster, they are usually the problem. Crowing problems usually fall under the noise problem, and you will usually find a special form for this called a "Noisy Bird Complaint Form" – try to avoid being a recipient of these forms if possible, and you should be fine.

***

Canada

In Canada, it seems that it is more complex. You cannot have a rooster in any city. They say no rooster as it is "noise in enclosed neighborhoods" being the issue. In Canada, the laws are made by each city and municipality.

As forward thinking as Canada is, most of their cities do not allow chickens in the backyard. Now, they don't go around looking for someone that is breaking the law, but if someone complains, they will go and check out the complaint.

***

United Kingdom

In the United Kingdom raising chickens in your backyard is regulated by DEFRA – this stands for Department for Environment, Food & Rural Affairs. You can keep chickens in your backyard anywhere in the United Kingdom. But, after the Avian Flu outbreak in 2005, they changed the regulations so that ANYONE that had more than 50 chickens is now regarded as a commercial chicken owner and must be registered with DEFRA.

If you are a member of a housing association, private property, council or a tenant, check in with that organization and make sure keeping chickens is okay. Sometimes they will have rules that you can't have them.

If you are a private home owner, read the deed to your house. As crazy as it may seem, there may be a clause in it somewhere that will say you can't have livestock on the place for so many years.

***

Rest of Europe

And the rest of Europe? Well, it encompasses so many countries that it is sort of complicated to give you one rule of thumb. There is a lot of information about poultry welfare, but it is mostly regarding large poultry producers and has no bearing on backyard chickens. You will need to look up your own country for its regulations as well as the knowledge from the urban area in which you live. In general though, roosters are frowned upon in urban areas.

***

So, bottom line, check with your local authorities and make sure you are within your rights before you get started and see if you must have a permit before you even get your chickens. That way you won't have any regrets down the road.

You never know what kind of neighbor you are going to have in town. You can have one that is so welcoming to chickens and glad to have them for their neighbor. Then there are others that will fuss about everything there is to do with a chicken. I can promise you that. They will even blame the weather on the chickens. However, offering them a steady batch of your girls' fresh eggs may work miracles!

Chickens are so amazing and so happy and can almost be a member of your family if you treat them like they are.

9. A Few Last Tips

KEY TAKEAWAY: This is the last chapter. I have added a few final tips and ideas here which hopefully will help you in your decision as to whether to take that step into a 'chicken' adventure in your backyard. I think it is worth the adventure. I know an 81 year old woman who manages a coop while still working 28 hours a week. If she can do it, I am sure you can too!

Some people may think we are a bit on the crazy side in the way we care for our chickens. But watching someone like my mother who is so dedicated to the health and welfare of her girls is quite endearing to me.

***

Giving Your Chickens Greens Year-Round

The chickens next door at my mother's home are never wanting for anything. They have a lovely chicken house, lots of healthy food, plenty of cracked corn and she intentionally plants them a large plot of turnip greens in the fall. The turnip greens can take frost well and are able to flourish good until there is a hard-killing frost.

Some people like to buy alfalfa hay for their chickens in the winter. But we are fortunate and have neighbors who run a catering business and they bring all their leftover salad fixings for the chickens in trade for eggs. This way, the girls have fresh greens year-round.

When the greens get large enough she starts pulling them a large fresh bunch every day and the girls know what she is coming out to do. They all run to the corner of the pen where they can watch her in the garden as her 'girls' are such a nosey bunch. Then they get so excited when they see her coming back. They all run back to the gate where she will enter to bring them their fresh greens. They get so excited when she lays them on the ground for them. They all talk amongst themselves and it sounds like they are talking about how good they taste.

I highly recommend you take the same approach!

***

Put on Aprons to Stop Picking

One summer, when her hens were molting and some had started their picking on each other, we ordered some chicken aprons for them. This also helps stop that picking.

The aprons are extremely cute. I picked the ones made of lightweight fabric so as not to make the girls hotter, but it covered the area where the feathers were missing well. It served two purposes. It stopped the picking and kept them from getting sunburned. You can order the pattern for the chicken aprons online and make them yourself out of small scraps of fabric.

***

How to Prevent Water From Freezing in the Bowl

Mom has found that other ways she keeps their water from freezing in extreme cold winters, which thankfully we have not had in quite a long time, is to use heated dog water bowls or metal cloth heating pads to sit under their watering cans. Another water bowl that works well is one that has marbles in it and when you plug it in it keeps the bowl jiggling slightly, causing the marbles to move and therefore, the water moves not letting it freeze.

***

Dealing with Roosters

The more time you can spend with your chickens, the better you will know them and vice versa. Before you know it, they will come running to you when you approach their pen as if to say, what exciting thing have you brought for us today?

The only aggression I have ever seen in the flock is when there were four roosters together. Three of them were pulled and kept in a separate pen to themselves until they died of old age. All had names just like the rest of the flock. They had some long spurs that could really do some damage to your legs if you had on short pants and they could sneak up on you when you were in the pen.

Sometimes, you must put up with a rooster if you want your eggs to be fertile and you want to hatch some eggs to enlarge your flock. If your flock is reaching two years old, you might want to add another dozen after a year and then a dozen every year after that. You will have some hens just fall over dead for no reason. You never know for sure what has happened. It could be that an egg ruptured inside of them, or they had a massive heart attack. They can be born with defects that do not show itself until later in life. You must take it in stride.

I had a gorgeous rooster who was on his post one morning and crowing his heart out. I mean, as roosters go, he was drop dead beautiful. That was a poor choice of words really. He was not that old, maybe 18 months but he had some bad habits, his favorite food was cracked corn. If he could, he would have made that his constant diet. That morning he was crowing away, in mid crow, he fell off his post, deader than a door knob. I kid you not. What a shock to all of us! I totally believe it was his diet. He would pick through everything and try to get only the cracked corn. I am sure he had a massive myocardial infarction! I read later that a constant diet of corn for a chicken causes large amounts of fat to build up around their hearts.

***

A Final Thought on Predators

Then you must be prepared if predators cause problems. Mother has had chickens for over seven years this time and never had trouble with predators until this past year. It has been very confusing why, out of the blue, they have started to come in, and all kinds at that, to have themselves a bite of lunch and midnight snacks off her hens. Most of them were her older hens, so I doubt they could fight anything off. Maybe they wanted a predator assisted suicide. We will never know.

***

I hope that all the odds and ends and tidbits I have shared throughout this book, as well as the personal stories, have helped you in some small way.

Chickens can be such a beautiful mix. Some reds, some blacks and some black and white striped ones. And they all sound so very happy talking in their pen.

Final Words

Thanks again for taking the time to read this book, *'Backyard Chickens: Join the Fun of Raising Chickens, Coop Building and Delicious Fresh Eggs (Hint: Keep Your Girls Happy!).'*

You should now have a good understanding of raising chickens in your backyard and be able to have your own fresh eggs and/or meat.

Having fresh eggs is just not comparable to anything else you might raise farm wise. The eggs you purchase at the store are supposed to be fresh, but it is easy to see when you put an egg in a pan to boil and it floats to the top, you know then that it is not a fresh egg as the sign boldly proclaims.

You know your eggs are fresh. You just went outside and pulled it from under your hen. She was happy to oblige as you have taken such loving care of her.

There are so many varieties of chickens, and if your children get interested, they can raise them for 4-H projects or to become 'show birds'. There are some elegant species out there, and there are some who love to be cuddled and babied as much as a new puppy or kitten.

Chickens serve as a good bonding experience and hobby with your children and grandchildren as well. If they start showing their prize chickens or just turn them into pets, the experience can be a unique and joyful one.

No matter what you decide in regard to chickens, I wish you well in your endeavors.

BONUS: CHICKEN JOKES!

These are 11 jokes from my popular book *'101 Hilarious Chicken Jokes & Riddles For Kids'*.

Enjoy!

***

1.

While mending fences out on the range, a very religious cowboy lost his favorite Bible. He was devastated!

Three weeks later, however, a brown chicken walked up to him, carrying that same Bible in its mouth.

The cowboy was astonished, he couldn't believe it! He took the precious book out of the chicken's mouth, thanked him, went on his knees and exclaimed: "It's a miracle!".

To which the chicken replied: "Not really. Your name is written inside the cover."

2.

Q: How do baby chickens dance?

A: Chick-to-chick!

3.

A chicken sits in a bar, sipping a whiskey.

A customer walks up to him and says, "Wow, it's not often that I see a chicken drinking bourbon here!"

To which the chicken replies: "Yeah, but that's hardly a surprise at these prices."

4.

Q: How did the eggs leave the highway?

A: They went through the eggs-it.

5.

One day, a man was driving on a country road when he looked out of the window and noticed a chicken running alongside his car. He was amazed to see the chicken keeping up with him: he was driving 40 mph! So, he accelerated to 50. But the chicken stayed right next to him. Even more astonished, he now sped up to 60 mph, but the chicken not only kept up with, it even passed him!

Then the man noticed something peculiar: the chicken had 3 legs. He decided to follow the chicken and finally ended up at a farm. When he got out of his car and looked around, he was even more shocked: all the chickens on this farm had three legs!

He approached the farmer and asked: "Why do all these chickens have 3 legs?"

The farmer replied: "Well, I figured: everybody likes chicken legs, right? So, I decided to breed a three-legged bird. I'm going to be a rich!"

Then the man asked him how the chicken legs tasted. Then farmer said, with a sad expression on his face: "I don't know, I haven't caught one yet..."

6.

Q: How do monsters like their eggs?

A: Terri-fried!

7.

Psychiatrist: What seems to be the problem?

Patient: I think I'm a chicken.

Psychiatrist: How long has this been going on?

Patient: Ever since I came out of my shell.

8.

Q: What does a chicken have in common with a band?

A. Drumsticks.

9.

Returning from the market, the farmer's son was dropped the crate of chickens his father had entrusted to him. The box broke open, and the chickens scurried off in different directions.

The boy panicked for a few seconds, but then quickly got back to his senses. Determined not to disappoint his father, he walked all over the neighborhood scooping up the wayward birds and returning them to the repaired crate.

Finally, the boy – reluctantly – returned home. He hoped he had found them all, but expected the worst.

He felt it was best to just speak up: "Pa, the chickens got loose," the boy confessed sadly, "but I managed to find all twelve of them."

"Well, you did really well son," the farmer replied. "You left with only seven this morning!"

10.

Q: Which side of a chicken has more feathers?

A. The outside.

11.

Q: Why shouldn't you tell an egg a joke?

A: Because it might crack up!

***

This is the end of this bonus chapter.

Want to continue reading?

Then get your copy of "101 Chicken Jokes" at your favorite bookstore!

Did You Like This Book?

If you enjoyed this book, I would like to ask you for a favor. Would you be kind enough to share your thoughts and post a review of this book? Just a few sentences would already be really helpful.

Your voice is important for this book to reach as many people as possible.

The more reviews this book gets, the more people will be able to find it and learn how they can keep chickens in their backyard!

***

IF YOU DID NOT LIKE THIS BOOK, THEN PLEASE TELL ME!

You can email me at **feedback@semsoli.com**, to share with me what you did not like.

Perhaps I can change it.

A book does not have to be stagnant, in today's world. With feedback from readers like yourself, I can improve the book. So, you can impact the quality of this book, and I welcome your feedback. Help make this book better for everyone!

Thank you again for reading this book and good luck with applying everything you have learned!

I'm rooting for you...

By The Same Author